BRIAN THE FIELD

Author: Ken Mackenzie
Illustrator: Ishika Sharma
Editor: Chris Stead

Written by Ken Mackenzie
Illustrations by Ishika Sharma
Editing and Design by Chris Stead

Published by Old Mate Media
www.oldmatemedia.com

Printed by Ingram Spark

Printed in Great Britain

First Edition

ISBN 978-1-925638-24-0

dedication

I dedicate all my stories to the children who are no longer with us, to all the children still fighting, and especially to those who have won the battle. If my stories made one of these children smile, then I am happy and my job is done.

Never far from my heart, my brother Colin Mackenzie, sadly missed

OUR CHARITY COMMITMENT

Supporting
children's hospice
SOUTH WEST
Registered Charity No. 1003314

Children's Hospice South West provides care for children who have life-limiting conditions and are not expected to live into adulthood, whilst also supporting their whole family. Old Mate Media is delighted to share that 25% of the profits from the sale of Brian The Field Mouse will be donated to the Children's Hospice to help them continue their essential work.

Find more information at www.chsw.org.uk

ABOUT THE AUTHOR

Ken Mackenzie lives in the United Kingdom: Torquay, South Devon to be exact. He is married and is the lucky grandfather to a huge tribe of 11 grandchildren. He loves nothing more than reading stories to the younger generation and taking them on a journey to a brighter, kinder world. He is devoted to the Children's Hospice, and writes his stories in honour of the children no longer with us.

Along, long time ago, there lived a little mouse called Brian. Brian was a field mouse who spent all his days in a field with his friends and family. Brian liked living in the field and he would spend all day playing with his brothers and sisters.

Every lunch time, they would all go into the big barn where the farmer kept the grain and they would eat and eat and eat. They'd eat until they felt they would burst.

It was a great life for some, but Brian was getting really frustrated. The farm simply wasn't big enough for such an adventurous mouse.

"I'm bored," he said to his friends one day. "I'm bored with doing the same thing every single day. I have decided to go travelling,"

"Travelling?" they all said in surprise.

"Yes," said Brian. "I am going out to see the world."

That made everybody very sad for nobody had ever left the field before. But Brian had made up his mind.

The next day, Brian began packing his bag with his favourite foods, grain and apples. Suddenly his best friend Alginod handed him a special note. On it were a list of things Brian might see and people he might meet if he ever had the courage to leave the field.

"Don't lose this as you will need it," said Alginod. "And there is one thing you must never forget...

ALWAYS WALK TO THE LEFT."

Brian thanked him, threw his bag over his shoulder, and said goodbye to his friends and family.

It was time to set off on his big adventure.

Brian crossed the farmer's field and walked up to the big wooden gates on the far side. He had never been that far before.

When he was little, his mummy and daddy told him, "never ever go to the big wooden gates, as that is where the world ends!"

So nobody ever went there.

Brian reached the far side of the field and stopped. He looked up to the top of the big wooden gates and they seemed to go all the way to the clouds.

"This is it," he said to himself and walked under the gates.

As soon as he walked through the gate he understood why nobody ever left the field.

In front of him was a giant monster. A monster with big bright eyes that lit up the whole field. It made a very loud noise, too, like a farmer's tractor. And it had lots of smoke pouring out of its back.

Brian was very frightened of the monster for he had never seen anything like this in his whole life.

Then he remembered something from Alginod's list. He quickly pulled it out and sure enough there it was. It was a picture of the exact same monster with the exact same bright eyes and a name.

It was called a car. Alginod had written a note next to it, saying, "never go near a car or you will be squashed and remember...

ALWAYS WALK TO THE LEFT."

Brian waited for the monster to pass and shortly after it was very quiet again.

"Thank goodness," thought Brian and then he set off once again on his big adventure.

He hadn't gone very far when a big moon began to rise. So Brian decided to find a nice dry, warm place to spend the night. He was just settling down when he heard something scary.

"CROAK, CROAK."

The sound made Brian jump and bump his head against a branch. Ouch! He very carefully leaned his head out of his nice warm sleeping spot to see if it was another monster.

And there it was right in front of him. The biggest pair of eyes he had ever seen.

"Who are you?" asked the eyes.

"I am Brian," answered our little mouse.

"And what is a Brian?" asked the eyes.

"I'm not a what," said Brian louder. "I'm a field mouse. What are you?

"I'm a wide-mouthed frog," said the eyes, and it pushed its face right up close to Brian's. He did have a wide mouth! "This is where I live and I've just finished my dinner."

"That's nice," Brian replied. "I have grain and apples for my dinner. What do wide-mouthed frogs eat?"

"Mice," said the frog.

Brian stood very still for a moment and then slowly walked backwards into his nice warm sleeping spot. He looked down at Alginod's note and it read, "don't ever talk to the wide-mouthed frog for it eats mice, and remember...

ALWAYS WALK TO THE LEFT."

Brian stayed in his nice warm sleeping spot the whole night, right up until the sun came out. He then very slowly and quietly peered outside to make sure the frog was gone.

Brian realised how lucky he was that the frog had already finished its dinner when they met last night. Otherwise, he might have ended up being a froggy dinner!

He picked up his bag and quickly ran through the hedge and away from the place where the wide-mouthed frog lived.

He walked and walked until he got really tired and very hungry. It turned out travelling was hard work. The world was a lot bigger than Brian's field.

After he stopped to eat some grain and apples, and enjoy a drink of water, Brian lay down to catch his breath. The sunshine felt warm on his face and soon he began to feel sleepy.

He was just falling into a deep sleep when he heard something.

"MEOW."

Brian opened one eye and there, standing over him, was a very big cat. It was fat and silver and had huge green eyes.

"Hello," Brian said nervously. "What are you?"

"Have you never seen a cat before?" asked the big silver cat. "What's in your bag?"

"It's my dinner," Brian replied, "grain and apples. Where are you going today?"

"Nowhere," said the big silver cat. "I'm looking for my lunch."

"Do you want some of my grain and apples?" asked Brian.

"Yuk!" said the big silver cat with a big wide grin, "I only eat mice!"

Oh no, not again, thought Brian. Then suddenly the farmer's dog jumped over the fence and ran into the field. The big silver cat's fur stood up like needles and off she ran as the dog gave chase.

"Phew," thought Brian, "I didn't like her very much."

He took out the note Alginod gave him and read the next entry: "never ever go near a cat, for they eat mice. And remember...

ALWAYS WALK TO THE LEFT."

Brian picked up his bag and continued on his adventure. He was very happy he had his note, but were there any more monsters out here?

"Hoot! Hoot!"

"Who are you and where are you going?" said a voice from up high.

"I am Brian," he answered, "and I'm going on a big adventure."

"Hmmm," said the voice as it swooped down to a lower branch. It was an owl. A big owl, with blue eyes and long sharp claws. "So you're called Brian? Well you look like a mouse to me!"

"Well," said Brian, "I... I... I am..."

He was about to say, "I am a mouse," when he remembered the note. He quickly pulled it out and read it. "Whatever you do," it said, "don't speak to the owl, for its favourite dinner is mouse."

"I... I... I'm a dog," Brian said, crossing his fingers behind his back.

"A dog?" said the owl. "You're a strange looking dog."

"Well, I'm a small farm dog," said Brian, "and I'm looking for my favourite lunch, yummy owl!

The owl wasn't going to wait around in case this very small farm dog ate him for lunch, so he flew away.

Brian was so very frightened by the big owl he decided maybe he needed a break from travelling. The big outside world was a very, very dangerous place to be. So he set off home remembering...

ALWAYS WALK TO THE LEFT.

Just as his best friend Alginod had told him. He walked and walked, until he spoted something he recognised. It was his field. It was his barn. It was his favourite grain and apples.

He was home! And all his friends and family came out to meet him.

That night he told his friend Alginod all about his adventure. He talked about how important the note had been to keeping him safe and that, just like Alginod had told him, Brian had...

ALWAYS WALKED TO THE LEFT

But Brian had a question for Alginod. "I understand your note about the car, and the frog, and the cat, and the owl, but why was it important to always walk to the left?

Alginod smiled. "I knew if you always walked to the left then no matter where your big adventure took you, in the end you would circle back home. You see, if you keep walking to the left, then you will go in a big circle, and end up right back where you started."

Brian laughed!

"It's good to go on a big adventure, but it's true that there's no place quite like home," Brian realised. "I must remember, when I'm out their on my own, to be very careful who I speak to and to never ever talk to strangers."

"And, of course, whatever happens, always come home."

Never go near a car
or you will squashed
and remember,
ALWAYS WALK TO THE LEFT

22

THE END

Turn your dreams of writing a book into a reality

Do you have a story you tell to your kids or grandkids all the time?
Are your friends and family always saying, "you should be an author?"

Old Mate Media specialises in helping indie authors
self-publish their ideas and own 100% of the copyright.
We will walk you down the path to being published.

Visit oldmatemedia.com to find out how we
can help you realise your dream.

Lightning Source UK Ltd.
Milton Keynes UK
UKHW052100041121
393388UK00005B/67